KEEPING THINGS COOL

THE STORY OF REFRIGERATION AND AIR CONDITIONING

Barbara Ford

WALKER AND COMPANY
NEW YORK

First published in the United States of America in 1986 by the Walker Publishing Company, Inc.

Published simultaneously in Canada by John Wiley & Sons Canada, Limited, Rexdale, Ontario.

Library of Congress Cataloging-in-Publication Data

Ford, Barbara.
  Keeping things cool.

  (Inventions that changed our lives)
  Includes index.
  Summary: Traces the history of our attempts to keep things cool, from the early use of fans and ice to modern refrigeration and air conditioning.
  1. Refrigeration and refrigerating machinery—Juvenile literature. 2. Air conditioning—Juvenile literature. [1. Refrigeration and refrigerating machinery. 2. Air conditioning] I. Title. II. Series.
TP492.F56 1986   621.5′6   85-26549

ISBN 0-8027-6635-8

ISBN 0-8027-6616-1 (reinf.)

Book Design by Teresa M. Carboni

Printed in the United States of America
10 9 8 7 6 5 4 3 2 1

CONTENTS

Egyptian King Amehopte and his mother are fanned
with feathers by a slave in this painting on the wall of
a tomb.—*Metropolitan Museum of Art*

1

Before
Refrigeration

ONE OF THE first references to keeping things cool comes from a Chinese poem written about 1100 B.C.:

*In the days of the second month, they hew out
the ice with harmonious blows;
And in the third month, they convey it
to the ice house.*

We don't know what an ancient Chinese ice house was like, but two other ancient civilizations, the Greeks and Romans, packed snow and ice from the mountains into cellars and covered it with an insulating material such as straw. Protected like this, the snow or ice might last all summer. In Rome, people bought snow in shops during the summer and used it to cool wine.

Ancient Egypt had no natural ice or snow, so the Egyptians cooled liquids another way. In the first century B.C., the Greek Protagoras described how they did it: "At night, they expose water to air in large earthen pitchers on the highest part of the house. Two slaves sprinkle the pitchers with water the whole night. By morning, the water has become so cold as not to require snow to cool it."

The Egyptians were taking advantage of two natural phenomena. When they sprinkled pitchers with water, this water evaporated and took heat from the water inside. You feel the same *cooling* effect if you sprinkle your hand and let it dry. When the Egyptians left a pitcher of water out on a clear night, radiation toward the cold sky formed ice crystals in the water. The same thing happens to an automobile parked outside on a clear night. Frost forms on the roof.

The fan appeared very early in history. The Egyptians fanned themselves (or had their slaves do it) with palm leaves or feathers. The Chinese used stiff paper, and the Japanese invented the folded fan. One of the first mechanical fans was invented by Leonardo da Vinci, who lived from 1452 to 1519. He cooled the Duchess of Milan's bedroom with a large revolving drum that blew air.

By the sixteenth century, people knew that you could produce temperatures below freezing

This fan worked with a bellows.—*Carrier Corporation*

by mixing certain chemicals with ice, snow or water. This is the principle of the hand-operated ice cream freezer. The ingredients of the ice cream are frozen by churning them in a container surrounded by the freezing mixture. Ice cream was popular in Europe in the sixteenth century, although it was probably made long before then.

High ceilings, big windows and porches kept homes like Longwood in Natchez, Mississippi, cool before air conditioning.—*Natchez Pilgrimage Tours*

One important way people kept cool was by building their houses to take advantage of natural cooling effects. In the hot, dry climates of Spain, North Africa and the southwestern United States, people erected thick-walled houses with interior courtyards. A fountain in the courtyard sprayed water, cooling the rooms around the courtyard by means of evaporation. The thick walls helped retain the coolness.

In the southeastern United States, the old mansions had thick walls as well as tall windows

4

Thomas Jefferson's ice house at Monticello.
—*Andy Johnson*

and big porches. The windows allowed the breezes to enter, and the porches kept out the sun.

Rural homes in the United States often had a springhouse, a small house with a tile or stone floor built around a spring. Spring water flowed through the house, cooling dairy products. In warm areas, wealthy landowners often had ice houses, too. Built partly underground, these ice houses stored ice cut from ponds or rivers. Thomas Jefferson had an ice house at his home, Monticello, and George Washington had one at Mount Vernon.

If the ice ran out, you picked up a chunk at the nearest warehouse.—*NY State Museum*

People took advantage of natural ice throughout history, but ice was used more widely in nineteenth-century United States than in any other time or place. Beginning around 1830, natural ice was a big business here. We used ice in homes, food markets, fishing boats and hotels and restaurants. Our brewing and meat-packing industries depended on enormous quantities of ice.

Why was natural ice so popular here? One reason was the invention of the icebox.

In 1803, a Maryland farmer, Thomas Moore, published a pamphlet describing his invention, the "ice box." An insulated box, it had a compartment at the top that held a big chunk of ice. Cold air from the ice moved downward, removing heat from the food as the ice melted. The water flowed into a pan at the bottom and was emptied by hand. By the early twentieth century, some iceboxes were large, handsome cabinets with such good insulation that the iceman came only once a week.

The natural ice used in the United States came from ponds, lakes and rivers in the northern states. New York City was one of the centers of the ice trade. Ice was cut on the Hudson River and nearby ponds and stored in huge, insulated warehouses. The warehouses kept the ice frozen until the warm months. In the summer, ice was loaded onto barges or wagons and delivered to

In rural areas, you cut your own ice.
—*Travelers Insurance Companies*

Men with handsaws cut ice in New York State.
—*NY State Museum*

distribution points in cities or towns. Icemen
with wagons delivered the ice to homes.

Some of our ice traveled long distances. In
1806, Frederick Tudor of Boston shipped ice all
the way from Boston to the Caribbean island of
Martinique. Soon Tudor, who became known as
the "Ice King," was shipping ice all over the
world. One of his employees, Nathanial J. Wyeth,
invented the ice plow, a device that made it easier to cut ice. It had runners with sharp teeth
that made deep grooves in the ice.

Just when natural ice was being used in
larger quantities than ever before, a new invention appeared—the refrigerating machine.

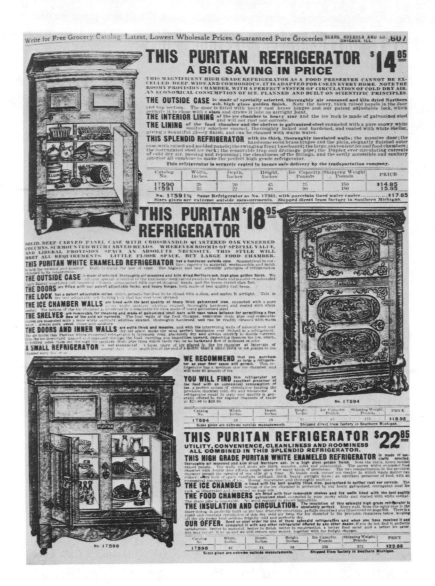

The Sears Roebuck catalog for 1910 offered iceboxes
(which they called refrigerators) for as little as $14.85.
Ice was put in the top of the cabinet.
—*Sears, Roebuck and Co.*

9

2

The Miracle Machine That Makes Ice

ON JULY 14, 1850, the French consul in the little Florida seaport of Apalachicola held a dinner party in honor of Bastille Day, a French national holiday.

"We have a special treat today," announced Monsieur Rosan to his guests. "Iced champagne!"

The guests looked pleased but surprised. The supply of natural ice stored in the town warehouse had been used up weeks before. And the ice shipment from Maine was overdue. But the champagne poured by the waiters was cold—almost as cold as ice.

"One of your countrymen has worked this miracle," said the consul. He raised his cham-

frigerator: heat flows only from a hot object to a cold one.

Carnot died young, in 1832, and other scientists expanded our knowledge of heat and cold. In the 1840s, a British scientist, James Joule, demonstrated Carnot's findings in the laboratory.

How much did John Gorrie, Jacob Perkins and Ferdinand Carré know about thermodynamics and the nature of liquids and gases under pressure? We don't know. They were practical men, and they obviously knew enough to build machines that used these laws. All the early refrigerators made ice successfully, although they did it in somewhat different ways.

Jacob Perkins's design worked something like Cullen's little laboratory experiment, but Perkins's invention was a practical apparatus. The working parts were a pump called the compressor, a valve (a device to regulate the flow of liquids and gases), a set of metal coils called the evaporator and a container, the condenser. Another container held water.

Perkins started with a special liquid. The valve reduced the pressure on this liquid as it flowed into the evaporator. The drop in pressure made the liquid cold. Heat from the container of water flowed to the cold liquid, just as Carnot's law says, and the water froze. Meanwhile, the special liquid evaporated because of the heat and low pressure.

Dr. John Gorrie.
—*Florida State Archives*

pagne glass, nodding at a stocky man with black hair. "Dr. John Gorrie!"

After dinner, Dr. Gorrie, a 47-year-old physician with a practice in Apalachicola, demonstrated his "miracle": a machine that produced ice cakes the size of bricks. Gorrie had been experimenting with ice machines for a half-dozen years, but it wasn't until 1850 that he completed a working model. He received a U.S. patent on it in 1851.

Gorrie's medical background led to his interest in ice machines. He believed that yellow fever, a serious health problem in the South at that time, could be prevented and treated by keeping indoor temperatures low. The physician devised

Dr. William Cullen.—*The Science Museum, London.*

a simple air-conditioning system, using ice made by his machine, and installed it in a hospital room.

Gorrie's ice machine was actually the second one to be patented. In 1834, Jacob Perkins, a New Englander living in Great Britain, received a British patent for a refrigerating machine, as the first refrigerators were called. Around 1860, a third machine was invented by a Frenchman, Ferdinand Carré.

Three different refrigerating machines in a 25-year period! Why did all three appear in such a short time, when none existed before? The time was ripe. For many years, scientists had been studying the nature of heat and cold. By the nineteenth century, this basic research enabled inventors to design refrigerators.

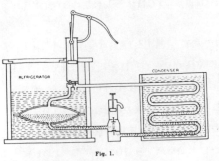

The patent for Jacob Perkins's refrigerating machine.

One of the first scientists to do basic research on cold was Scottish physician Dr. William Cullen, who lived from 1710 to 1790. Cullen put a container of water in an airtight jar and removed most of the air from the jar with a pump. The removal of the air reduced the pressure on the water. Some froze and some changed to a gas, or evaporated. Cullen had made artificial ice.

Cullen's little experiment demonstrated *a natural law*: water freezes at low pressures. This same law is used in a different way in today's refrigerators.

In 1825, a young French engineer, Sadi Carnot, wrote a scientific paper in which he laid the foundations for the science of thermodynamics. Thermodynamics describes the nature of heat and work and the transfer of energy from one to the other. Part of Carnot's Second Law of Thermodynamics tells us what happens inside a re-

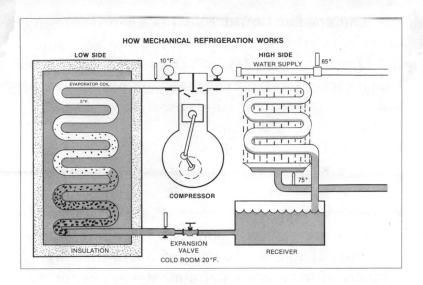

A modern compressor refrigerator works essentially like Jacob Perkins's 1834 machine. The liquid refrigerant absorbs heat and changes to a gas in the evaporator (left). Then the gas gives off heat and changes back to a liquid in the condenser (right). The compressor (center) keeps the refrigerant moving and, with the valve (center), regulates pressure. Note that the evaporator and condenser are coiled tubes.—*International Institute of Ammonia Refrigeration*

This part of the invention worked much like Cullen's laboratory experiment. But Perkins took refrigeration one step further. The compressor raised the pressure of the gas, which was now warm, and pumped it into the condenser. There it was cooled by air. The high pressure and the cooling made the gas change back into a liquid, or condense.

The liquid flowed back into the evaporator, where the whole cycle started over again.

The special liquid Perkins used in his machine was ethyl ether. We call these refrigerating liquids "refrigerants." They are the real key to a practical refrigerator. Their properties allow evaporation and condensation to take place within the refrigerator.

John Gorrie's machine used air as a refrigerant. Pressure on the air was reduced, which allowed it to become cold. The cold air chilled a mixture of salt and water which was used to freeze water. Carré's machine was the most complex. It had a number of parts in which a mixture of water and a chemical was evaporated and condensed several times.

In spite of its complex design, Carré's machine was the first one made commercially. During the U.S. Civil War (1860-1865), several Carré machines were shipped through the blockade that the U.S. Navy put around Southern ports. One of these machines was used in a hospital that cared for wounded Confederate soldiers.

In the 1870s, Carré machines chilled carafes of water and wine in restaurants in Paris. People could buy these *carafes frappés* to take home.

Jacob Perkins's machine and John Gorrie's machine were not successful during their lifetimes. Perkins, who was 68 when he invented his refrigerator, apparently never tried to produce it on a commercial basis. Gorrie tried to raise money to develop his machine, but he failed.

Dr. John Gorrie's statue
in the U.S. Congress.
—*Florida State Archives*

When he died, just four years after receiving his patent in 1851, he was almost bankrupt.

But fame came to Gorrie after death. Others improved his invention, and the "cold air" machine was widely used in the last quarter of the nineteenth century for refrigerating meat aboard ships. Today this type of refrigeration cools the cabins on jet aircraft. In 1914, the state of Florida recognized Gorrie's contributions to refrigeration and air conditioning by erecting a marble statue of him in Statuary Hall in the U.S. Congress in Washington, D.C.

3

Frozen Meat and Fresh Vegetables

THE EARLIEST REFRIGERATORS were ice-makers. After the Civil War ended in the United States, a number of Ferdinand Carré's machines were imported to this country. They were used in factories to make ice, which was sold to homes and businesses. By 1900, every region of the United States had ice factories. Artificial ice was an even bigger business than natural ice.

Although natural ice was still used, each year less and less of it was harvested. By the 1920s, natural ice was no longer a big business.

The first ice machines were based on Carré's design, but soon improved versions of Jacob Perkins's compressor machine entered the field. Both machines were powered by steam engines during this period. In the 1860s, Carré introduced

This huge 1895 compressor refrigerator made ice. The manufacturer, the Frick Company, still makes industrial refrigerators.—*The Frick Company*

the use of ammonia as a refrigerant. It proved to be a good choice for compressor machines, too. Ammonia condenses at pressures easy to pro-

A refrigerating machine in a brewery in 1883.
—*Western Brewer/U.S. Brewers Ass'n*

duce, and it works very efficiently. It's still widely used today in industrial refrigerators.

When artificial ice first became available, breweries and meat packers were the biggest customers. Breweries used ice to keep beer cool while it was being made. Anheuser-Busch Companies in St. Louis, Missouri, used 40,000 tons of ice each year to brew its beer! Breweries had special ice houses to store the ice. Meat packers used ice to preserve meat before it was sold to customers. Ice created problems for these big users, though. It was messy, it required lots of space and it produced high humidity.

Around 1860, a few breweries in Europe and Australia found a better way. They installed compressor refrigerators and circulated cold brine (a mixture of salt and a liquid) through pipes in the ceilings of their plants. It worked so well that by 1890 practically all breweries had switched from ice to refrigeration. Meat-packing plants made the same switch.

Refrigeration soon solved another problem. South America, Australia and New Zealand raise large numbers of animals for meat. Europe, particularly the island nation of Great Britain, needs meat to feed its large population. In the nineteenth century, it took at least two months for a sailing ship to get from South America, Aus-

Refrigeration pipes cool cellar where beer is stored in this 1883 scene.—*Western Brewer/U.S. Brewers Ass'n*

The *Dunedin* carried the first cargo of frozen mutton
from New Zealand to England in 1882.
—*National Maritime Museum, London*

tralia or New Zealand to Europe. Fresh meat
can't be preserved that long, and ice can't make
it cold enough to freeze.

In 1876, Frenchman Charles Tellier installed three compressor refrigerators on the
ship *Frigorifique*, which means refrigerator. A
three-masted sailing ship, the *Frigorifique* had a
small steam engine which pushed it along at a
top speed of seven knots. Loaded with the frozen
carcasses of twelve sheep, ten cows, and two
calves, the ship set sail from France for Buenos
Aires, Argentina.

The voyage took 106 days. After the ship
made port, the meat was defrosted and tasted. It
was edible!

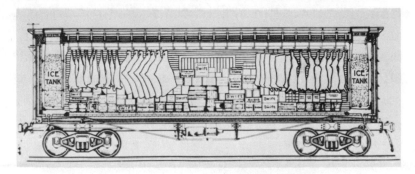

Swift Packing Company used freight cars cooled with ice to carry meat. The ice was in compartments at ends of cars.—*Swift Independent Packing Co.*

The *Frigorifique's* voyage was the beginning of the busy meat trade that still carries frozen meat between producers on one side of the globe and consumers on the other. Most of the early refrigerated sailing ships used cold-air machines to keep meat frozen. John Gorrie's invention used no refrigerant, so it was considered safer on ships than other refrigerators were.

The United States played only a minor role in the overseas meat trade because we raise most of our own meat. But refrigeration is very important in shipping food within our large country. After the Civil War, railway cars refrigerated with ice—natural and artificial—were used to haul perishable foods. One popular design had open compartments at each end of the car. Ice was poured in through hatches on the roof.

Iced railway cars worked well for meat, which doesn't have to travel long distances in the

United States. But most of our produce—fruits and vegetables—comes from the South and West, far from markets in the big cities of the East and Midwest. In the early days of refrigeration, a railway car took twelve days to cross the continent and a week to go from south to north. The ice soon melted.

In the 1890s, when ice factories were in all parts of the nation, the solution was found: the re-icing station. Wooden platforms the height of a railroad car were built right next to the tracks at points along the route. An ice factory was located near the station. When a train with fresh produce neared a re-icing station, it telegraphed ahead.

Workers climbed on the platforms and filled a chute with ice. When the train arrived, they opened the hatches and poured in the ice. The whole operation took only a few minutes.

When produce reached its destination, it was usually stored in a cold-storage warehouse. These were huge, multi-storied buildings cooled with some form of compression machine. Delicate produce such as lettuce was moved out of the warehouse quickly, but apples and other hardy items might stay there for months. Cold-storage warehouses also stored dairy products.

One final step was added to the "cold chain," the journey that took produce from field to consumer. Scientists in the U.S. Department of Ag-

At re-icing stations, workers pushed ice through hatches
in roofs of freight cars.—*National Archives*

riculture found that if produce was cooled by
refrigeration *before* it was loaded on a refrigerated rail car, it kept fresh much longer. Refrigerated facilities were built near growing areas to
cool produce as soon as it was brought in from
the field.

The cold chain works so well with produce
that people living far from growing areas are
able to enjoy fresh fruits and vegetables even in
winter. It changed the kinds of produce we eat,
too. Before refrigeration, only the hardiest kinds
of produce were available; today we can enjoy
delicate varieties.

An artificial ice skating rink in Paris in 1890.
—*Victorian Inventions, John Murray*

Most of the early refrigerators cooled food, but soon refrigerators were used for other purposes. In 1876, the same year Charles Tellier shipped frozen meat from France to Argentina, the first indoor ice rink was built in London, England. Manufacturers of such products as chemicals, rubber and photographic film were early users of refrigeration, too. In the 1890s, engineers who built tunnels and dams began to use refrigeration to freeze soil.

One of the biggest engineering projects involving refrigeration was Grand Coulee Dam in Washington State. In 1936, the U.S. Bureau of Reclamation dam builders had trouble with earth sliding into the excavation. They built a refrig-

Ice dam holds back slippery soil at site of Grand Coulee
Dam.—*U.S. Bureau of Reclamation*

eration plant and buried miles of cooling tubes in
the earth. The system produced an ice dam 100
feet wide, 40 feet high and 20 feet thick to hold
the soil in place.

In 1895, German scientist Carl von Linde made another breakthrough in refrigeration. He liquified air in a special refrigerating system with a series of compressors. The very cold liquid air that resulted was used in a chemical plant. From then on, scientists found more and more uses for the very cold liquids.

In the nineteenth century, all three types of refrigeration were popular. But by the end of World War I in 1918, the compressor machine, now powered by electricity, had moved ahead. Today most refrigerators, including the ones in our kitchen, are compressor machines. This machine has proved to be the simplest to operate, the most reliable and, for most purposes, the most economical.

4

Willis Carrier's Scientific Air Conditioner

WILLIS H. CARRIER walked back and forth on a foggy railway platform. It was the fall of 1902. Carrier, 26 years old, was an engineer employed by the Buffalo Forge Company, New York. He had just installed one of the world's first air-conditioning systems in a printing plant in Brooklyn, New York.

The customer, the Sackett-Wilhelms Company, was happy with the system, but Carrier wasn't. There must be, he thought, a more reliable way to control moisture in the air.

As he paced the platform, Carrier stared at the fog. Fog! His mind raced. The temperature was in the thirties and there was little moisture in the air, even though it was saturated with water. The air had to be saturated because fog is

Willis H. Carrier.
—*Carrier Corporation*

composed of tiny droplets of water. What kept the moisture content of the air—its humidity—so low? Suddenly the young engineer had a flash of inspiration.

Low temperatures, he realized, keep humidity low. If he could control the temperature of the water used in an air conditioner, he could control the temperature of the air and its humidity. When dry air is needed, he thought, I'll use a spray of cold water to lower the temperature of the air. The water will be cooled by a refrigerator. When moist air is needed, I'll heat the water.

The system Carrier dreamed up that night in 1902 soon became the basis for Carrier air conditioners installed all over the world. Carrier was

Willis Carrier installed his first air conditioner in a
Brooklyn printing firm, Sackett-Wilhelm, in 1902.
—*Carrier Corporation*

able to guarantee his customers whatever humidity and temperature they needed to make a
wide range of products. His first customers included textile mills and tobacco factories.

The Carrier air conditioner worked much like
a refrigerator, because an air conditioner is
really a kind of refrigerator. In addition to the
compressor and condenser and evaporator, the
air conditioner has fans and filters. The filters
clean the air, and fans blow the clean air over the
refrigerating system and then into the area
being cooled.

In Carrier's new machine, the extra touch was the cold water spray that lowered the temperature of the air. The water was chilled by refrigeration.

In large industrial and commercial machines like the ones Carrier designed, most of the machinery is in a separate room. Large pipes, called ducts, carry chilled air to the other rooms. The installations usually have a "cooling tower" on the roof. In the tower, cool water removes heat from warm water produced by the system. When the warm water is cooled, it goes back into the system.

Willis Carrier didn't make the first air conditioner. John Gorrie devised a simple room air conditioner back in the 1850s. A block of ice in a container hung from the ceiling, beneath a pipe that led through the ceiling. Air circulated through the pipe, over the ice and out through a hole at the bottom of the wall. Around the turn of the century, engineers installed more sophisticated systems in large buildings.

The term air conditioning was first used by one of its pioneers, Stuart W. Cramer. He air conditioned a number of textile mills in the southern states.

These early air-conditioning systems worked, but not very efficiently. Before Carrier, no one understood how to control the humidity in the air. When we feel uncomfortably hot, it's usually because both the temperature and the humidity are high. Reduce the temperature without reducing the humidity and we still feel uncomfortable. Carrier's air conditioner controlled temperature and humidity. It was the first scientific air conditioner.

Willis Carrier's new air conditioner brought so many new customers to the Buffalo Forge Company that the firm made him head of a separate company, Carrier Air Conditioning. In 1911, Carrier read a scientific paper he had written to a meeting of the American Society of Mechanical Engineers. It described how an engineer could use a formula to figure the air-conditioning needs of any system.

Soon the chart Carrier had made based on his formula was being printed in engineering textbooks all over the world. Students still study it today.

Just when Carrier's genius was being recognized, World War I broke out. In 1915, the Buffalo Forge Company decided to close Carrier Air Conditioning. Carrier and six other employees pooled their money and formed their own company, the Carrier Corporation. Carrier was president. The first office had two desks, a drafting board, a stool and a few files. Visitors sat in two wicker chairs.

By the end of 1915, orders were coming in fast, some from industries that had never used air conditioning before. The new firm was a success.

In the 1920s, Carrier invented more efficient refrigeration machinery to be used with his air conditioner. This invention and safe new refrigerants made it possible to use air conditioning in other fields. Until then, almost all air conditioners had been used to make products, not to keep people comfortable. The first breakthrough into comfort air conditioning by the Carrier Corporation was at the J. L. Hudson Department Store in Detroit.

After the success of the Hudson machine, several Texas movie theaters installed Carriers. Ticket sales boomed during the hot, humid Texas summer. The Rivoli, a large movie theater in New York City, heard about the success of the Texas theaters and bought a Carrier machine in 1924. Carrier realized the New York theater was a real test. If his improved air conditioner could make it there, it could make it anywhere!

The Rivoli was closed while the machine was being installed. The reopening was scheduled for Memorial Day, 1925. Signs saying "Cooled by Refrigeration" hung outside the theater. Inside, Carrier and his engineers worked up to the last minute. The air conditioning wasn't turned on until the audience of 2,000 was seated. Watching from the wings, Carrier saw 2,000 people fanning themselves with paper fans.

Then, as cold air filled the theater, the fans dropped into laps. After the movie, Carrier saw

The people liked the Rivoli Theater air conditioning.
—*Carrier Corporation*

Hollywood producer Adolph Zukor in the lobby. "People are going to like it," said Zukor.

And they did. Soon Carrier air conditioners were being used in theaters, restaurants, hotels, department stores, hospitals and office buildings all over the country. In 1928 and 1929, the U.S. Congress used Carrier equipment to cool its chambers. In 1931, the first air conditioner was installed in a railway car, the dining car of the *Martha Washington*, which ran between New York and Washington.

One of Carrier's final achievements was a special air-conditioning system for skyscrapers.

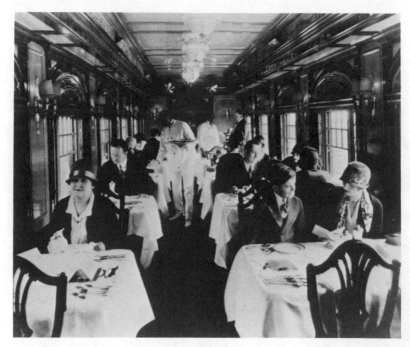

The dining room of the *Martha Washington* was air conditioned in 1931.—*Baltimore & Ohio RR,* —*Carrier Corporation*

Floor space in skyscrapers is tight, and large air conditioners need space for the ducts that carry cooled air to individual rooms. Carrier's solution was to reduce the size of the ducts and increase the velocity of the air. The air flows to small cabinets in each room where room temperatures can be selected.

The United Nations Building, which was completed in 1952, was one of the best-known buildings to use the new system.

The United Nations Building was one of the first
skyscrapers to be cooled by Willis Carrier's new
air-conditioning system.—*UN Photo/Milton Grant*

By the time Willis Carrier died in 1950, comfort air conditioning had become as important as air conditioning used in making products. Today more air conditioning is used to cool people than for any other purpose.

5

Refrigeration in the Home

INDUSTRY, BIG commercial enterprises such as department stores, and even some small businesses had refrigeration or air conditioning by the 1920s. But not homes. Homes depended on the visits of the iceman until well into the 1920s. Some had iceboxes until much later. Home air conditioning wasn't widely used until the 1950s.

Why did it take so long to get refrigerators and air conditioners into the home?

The early refrigerators weren't suitable for home use. Not only were they large and noisy, but they used too much power and were difficult to run. Skilled operators were needed to keep them working. All this was expensive, of course. The refrigerants used were not safe for the home, either. Early air conditioners had many of the same problems.

By the early years of the twentieth century, refrigerators had become smaller, more efficient and easier to operate. Electricity, not steam power, became available to run them. But refrigerators still left a lot to be desired. For one thing, they came in several parts. The cabinet that held the food was in the kitchen, while the compressor and much of the machinery were in the basement.

Between 1920 and 1930, three developments made the modern home refrigerator possible: a self-contained refrigerator with all parts in one machine, automatic controls and a safe refrigerant. The refrigerant was Freon. Freon was discovered in the 1890s, but it wasn't until 1930 that refrigeration experts realized its suitability for home use. Today Freon, a Du Pont Company product, and similar refrigerants are used in almost all home refrigerators.

In 1925, one of the first self-contained refrigerators was introduced by the Kelvinator Company. It took its name from British scientist Lord Kelvin, who did some of the early research on refrigeration. The "Kelvinator" had a thermostat that regulated its temperature. The Kelvinator Company had been making iceboxes for years, and its first refrigerators could be operated either with ice or with electricity.

A few years later, the General Electric Company came out with the Monitor Top refrigera-

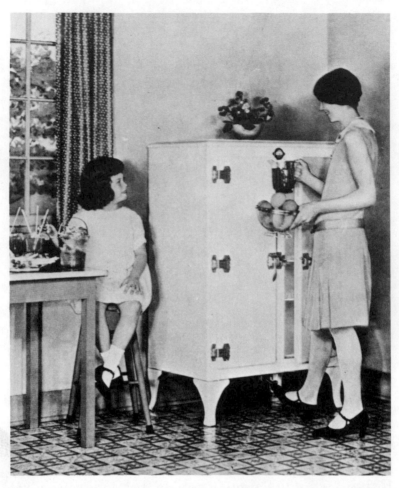

The Kelvinator was one of the first home refrigerators.
—*Kelvinator Appliance Company*

tor. The Monitor Top stood on four legs, like the
Kelvinator, but it had a big bump on top—the
compressor. When you opened the door, you saw,
above the top shelf, a tiny freezer surrounded by
the coils of the evaporator. The freezer could
freeze only a few trays of ice.

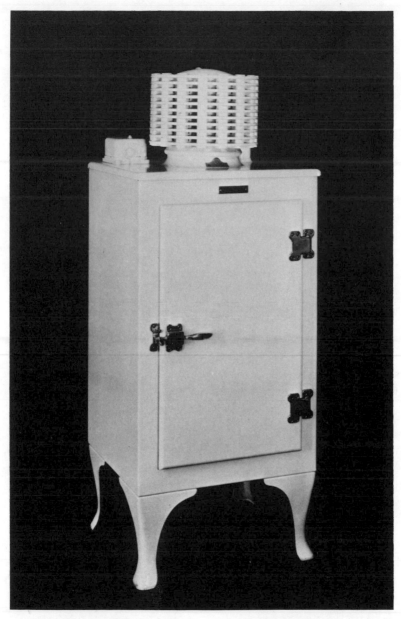

The Monitor Top Refrigerator had the compressor on top.—*General Electric Company*

GE's advertising reflected people's concerns in those days. "All the mechanism is here," says one early ad. "There's no machinery underneath, none in the basement!" Another ad reads: "The best thing about the goodies that come out of the General Electric Refrigerator is that they're always healthfully fresh." The ads also stressed the machines' quiet, automatic operation.

Sales of the new home refrigerators soared. In 1929, as many refrigerators as iceboxes were in American homes; by 1935, there were far more refrigerators. Improvements came quickly. In the 1940s, the two-part refrigerator with a large freezer appeared. In these new machines, the machinery was in the back, sides and underneath, where it is today. Separate home freezers came out at about the same time as the two-part refrigerators.

Freezers came along just as frozen food was becoming available. An American inventor, Clarence Birdseye, made frozen food an item on the American shopper's food list. His success was due to two things: his double-belt freezer and his quick freezing system. His freezer had two belts between which the food was held. Cold brine was sprayed on the belts. The quick freezing process resulted in smaller ice crystals and less damage to the food.

Birdseye was very successful with frozen fish

An early display case for Birds Eye frozen foods.
—*General Foods Archives*

and meat, but when he turned to vegetables, problems developed. The peas and other items with which he experimented lost taste and color. Finally Birdseye discovered that if he cooked the vegetables for a short time before freezing, the problems disappeared. Cooking killed enzymes that led to the loss of taste and color.

Now all of Birdseye's frozen food looked and tasted good, but he had trouble selling it. In 1929, the inventor sold his company to the Postum Company, which soon became General Foods. Birdseye continued to do research for General Foods. General Foods launched a test campaign in Springfield, Massachusetts, to interest people in frozen food. The headline for an ad in the local newspaper on March 6, 1930, read:

For the First Time Anywhere!
The most revolutionary idea in the history of food will be revealed in Springfield today

When people came to the stores mentioned in the ad, they found frozen food cases stocked with low-priced meat, fish and vegetables, all with the name "Birds Eye." People bought the frozen foods in Springfield, and soon they were buying them all over the country. By 1937, Birds Eye was making a profit. It's still the best-known name in frozen foods.

Birds Eye didn't invent the frozen dinner. The credit for that goes to C.A. Swanson & Sons, a small company in Omaha, Nebraska. Its first frozen dinner, which featured chicken, appeared in 1954 and cost about one dollar. Swanson was later bought by the Campbell Company, which now makes a wide variety of Swanson frozen dinners.

The air conditioner took even longer to reach the home. Several companies, including Carrier, tried unsuccessfully to sell home air conditioners in the 1930s. The units cost more than people wanted to pay. Just before World War II, the Westinghouse Corporation came out with a $149.50 model. After the war, it sold in large

The 1930 ad that sold Birds Eye frozen foods.
—*General Foods Archives*

numbers. By the 1950s, a number of companies were selling room air conditioners.

Room air conditioners have all their machinery within a box placed in a window or a wall.

Today the Campbell Company makes a wide variety of
Swanson frozen dinners.—*The Campbell Company*

A fan blows indoor air over refrigerating coils to
cool it. Then the fan blows the cooled air directly
into the room. The compressor, condenser, and
one fan are outdoors; the filters, an evaporator
and another fan are inside.

Home air conditioners and refrigerators have
had a major effect on people's lives, especially in
this country. We have more refrigerators and air
conditioners than any other nation. Refrigerators make it easier for the homemaker to buy and
prepare food. Before refrigeration, people had to
shop often because food didn't keep as long. And

since there were no frozen foods, everything had to be prepared "from scratch," which took longer. Refrigerators give the homemaker, who is often a woman, more time to pursue interests outside the kitchen.

Home and office air conditioners have had an important effect on where people live in the United States. Before air conditioning, such hot, humid southern cities as Houston, Texas, attracted few people from northern states. These cities were very uncomfortable during the summer. After comfort air conditioning became widespread, people began to move south in large numbers. Today Houston is our fourth largest city. It has more air conditioning than any other city in the world.

The year 1953 was a special year in the history of refrigeration. The last icebox in the United States was manufactured that year, 150 years after the icebox was invented. If you look in the yellow pages of the telephone book, you'll see that ice is still a business. It's used on fishing vessels, in fish stores, and for other special purposes. But today we all depend on refrigerators and air conditioners to keep things cool.

6

New Ways to Keep Things Cool

WHEN THE STATUE of Liberty was being cleaned for its 100th birthday in 1986, several problems arose. One concerned the copper statue's interior surface, which was covered with seven coats of paint. How could it be removed without damaging the statue or producing harmful fumes inside the statue? Frances Gale of Columbia University, an art expert who worked on the cleaning project, decided to try liquid nitrogen.

Liquid nitrogen is a very cold liquid made from a gas in a special refrigerating apparatus. It has a temperature of 320 degrees below zero. Union Carbide's Linde Division furnished enough liquid nitrogen to make a test. When workers sprayed it on a small area with a special tool, the

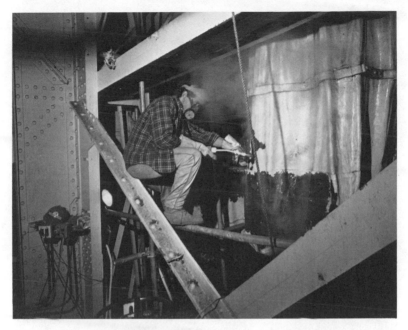

Worker cleans interior of Statue of Liberty with liquid nitrogen.—*Union Carbide Corporation, Linde Division*

cold liquid cracked the paint so it could be blown away. Union Carbide donated enough liquid nitrogen to do the whole interior, an area of about 11,000 square feet.

"Liquid nitrogen made it an easy clean-up job," says Frances Gale.

For many years, the production of cold liquids took place mostly in research laboratories that study cryogenics, the science of very low temperatures. But today we've learned how to put these liquids to work. Liquid oxygen is used in making steel. Liquid hydrogen makes a good rocket propellant. Liquid nitrogen not only re-

Dr. Clarence Gibbs of the National Institutes of Health selects blood sample from collection stored in refrigerated room.—*National Institutes of Health*

moves paint but is used in a special medical instrument that cuts human tissue. Liquid natural gas—LNG for short—is an important fuel. It is liquified to reduce its volume for shipping.

One of the most exciting properties of supercool liquids is superconductivity. Discovered in 1911 by Dutch scientist Kamerlingh Onnes, superconductivity means that at very cold temperatures—below 450 degrees Farenheit—there is no electrical resistance. Superconductivity has

Astronaut wears air-conditioned underwear beneath
space suit. Refrigeration unit in spacesuit supplies
underwear with cooled liquid.—*NASA*

already been used to improve reception from
communications satellites and in a special mag-
net, the "cryomagnet," used in research laboratories.

Louisiana Superdome in New Orleans needs 9,000 tons of air conditioning. Football and baseball are played there.—*Jack Beech, Louisiana Superdome*

Using very cold liquids is just one of the new ways we put to work refrigeration and air conditioning today. Laboratories chill or freeze organs, sperm, vaccines and viruses until they're needed. Movie studios and ski resorts make artificial snow with machines. Libraries preserve valuable books in air-conditioned surroundings. Large computers need air conditioning to work efficiently. Spacecraft and even spacesuits are air conditioned.

But in spite of all the new ways we've found to use refrigeration and air conditioning, most of our cooling is still used for traditional purposes.

52

We still use most of our refrigeration to preserve food, and we still use most of our air conditioning to keep poeple comfortable. But there have been big changes in the ways these inventions work for us.

Today most cold-storage facilities are on one floor, so automated equipment can be used to move items. Gases such as carbon dioxide are often pumped into produce storage rooms to lengthen the life of fruits and vegetables. Refrigerated trucks, not trains, carry most of our produce. Refrigerated trains and trucks both use mechanical equipment, not ice.

Some refrigerated trucks are carried, piggyback style, on flat-bed railroad cars. When the

The New York Public Library air conditions its book stacks to preserve the books.—*Anne Day*

A refrigerated truck is hoisted aboard Santa Fe Railway
car.—*Santa Fe Railway*

train reaches its destination, the truck drives
away.

In our food stores, the high, closed refrigerated meat case has been replaced by the low,
open case in which shoppers can help themselves. Delicatessen items, frozen foods and dairy
products all have their own self-service refrigerated cases. All these cases carry a much wider
variety of items.

Our home refrigerators and air conditioners
are different, too. The latest General Electric refrigerators offer such special features as an elec-
tronic diagnostic system that tells you if the

Modern refrigerators like this GE model have refreshment centers and ice and cold-water dispensers.
—*General Electric Company*

machine is working properly and a refreshment center built right into the door. The center has its own little door so you can remove items without opening the main door and wasting energy.

Many home air conditioners now cool the whole house, not just a single room. The whole-house air conditioner sends cool air through ducts to individual rooms. It does a better job of controlling temperatures and humidity than room models, and it's quieter, too. Some homes have a heat pump, a machine that heats in winter and cools in summer.

New air conditioners use less energy than older models. Manufacturers reduced the speed of the fans and increased the size of the evaporator coils to make them more efficient.

In the future, say energy experts, we may be able to use the power of the sun to operate refrigerators and air conditioners. Tomorrow's solar appliances will probably be powered by photovoltaic cells, the same kind of cell used on spacecraft. In fact, these cells already run appliances in some remote areas. Photovoltaic cells are still too expensive for most homes but, if the manufacturing costs are reduced, the sun will help us keep cool.

Cryogenics may play a part in filling our energy needs, too. Electricity could someday be stored and carried by means of superconducting cables. Superconductivity may even enable us to produce power by means of nuclear fusion. Very hot plasma is produced in fusion, and it might be possible to contain this plasma by a giant-sized

Someday photovoltaic concentrators like this one at Sandia Laboratories may power refrigerators and air conditioners.—*Sandia Laboratories*

version of the cryomagnet used in research laboratories today.

It looks as though efficient refrigerators and air conditioners will keep on doing more and more jobs for us in the future—some of which are not even dreamed of yet.

Acknowledgments

MY SPECIAL THANKS to Dr. Eugene Stamper of the New Jersey Institute of Technology, who furnished valuable help on the technical aspects of refrigeration. Thanks also to those who went beyond the line of duty in providing information and/or illustrations: Henry Bonar, Bonar Engineering, Inc.; Jack Bulleit, Frick Company; Franklyn Carr and Howard N. Barr, Chessie System Railroads; the Carrier Corporation; W. D. Dillman, Swift Independent Packing Company; Timothy A. Fausch, *Solar Engineering & Contracting*; Alan R. Fletcher, The Travelers Companies; Lucia S. Goodwin, Monticello; Ruth Greenamyer, General Foods Corporation; Jane E. Hazen, *Air Conditioning, Heating and Refrigeration News*; Marvin W. Heininger, Air Conditioning and Refrigeration Institute; Tom Hughes, U.S. Bureau of Reclamation; Patrick A. King, U.S. Brewers Association, Inc.; Gary R. Mormino, Florida Historical Society; Alice Peterson, Kelvinator Appliance Company; Moira J. Skislock, Union Carbide Corporation; Lenore Swoiskin, Sears; Craig S. Williams, New York State Museum.

Carrier Corporation

Index